HOW SENIORS CAN CARE FOR THEIR BEARDED DRAGONS

An In-Depth Guide on Meeting the Needs of Your Bearded Dragon: Learn the Essentials of Proper Care to Ensure a Happy, Healthy, and Well-Fed Companion

Lopez Tommy

i

DISCLAIMER

This book is intended to offer competent and dependable information on the topic at hand. The opinions stated in this publication, however, are solely those of the author and should not be construed as expert teaching or professional advice. The reader is personally accountable for his or her conduct. The author expressly disclaims any duty or liability arising from the purchaser's or reader's use or use of the contents of this book. The purchaser or reader accepts full responsibility for his or her conduct.

Every right is retained. Except for short excerpts contained in critical reviews and certain other noncommercial uses allowed by copyright law, no section of this book may be copied, duplicated, or reproduced in any form or medium, whether photocopying, recording, or other electronic or mechanical techniques, without the prior written consent of the publisher. For permission requests, please contact the publisher at the address shown below.

TABLE OF CONTENTS

INTRODUCTION

Bearded dragons have gained popularity as beloved pets among seniors, known for their gentle nature and appealing personalities. These intriguing reptiles have captured the hearts of many older individuals, owing to their charming disposition and simple care requirements. This chapter aims to explore why bearded dragons make excellent companions for seniors, highlighting the various benefits that come with caring for them in later years. Additionally, it will provide a comprehensive overview of how this book can guide seniors in taking care of their bearded dragons, ensuring a rewarding and fulfilling experience for both the seniors and their scaly friends.

Why Bearded Dragons Make Great Companions for Seniors

Bearded dragons have a natural ability to form strong bonds with their human caregivers, making them ideal companions for seniors seeking affection and companionship. Their calm and gentle demeanor creates an atmosphere of peace and comfort, offering seniors a sense of emotional stability and companionship that can be especially valuable in later stages of life. Unlike some other pets, bearded dragons do not demand vigorous physical activity, making them suitable for seniors with limited mobility or energy. The soothing presence of these reptiles can alleviate feelings of loneliness and seclusion, fostering a nurturing environment that promotes mental well-being and emotional balance.

Furthermore, the interactive nature of bearded dragons allows seniors to engage in meaningful and stimulating activities, providing a sense of mental agility and sharpness, potentially reducing the risk of cognitive decline and age-related neurological conditions.

How This Book Can Help Seniors Take Care of Their Bearded Dragons

This comprehensive guide has been thoughtfully created to provide seniors with a deep understanding of the intricacies involved in caring for bearded dragons. It offers valuable insights, practical tips, and expert advice to ensure a gratifying and successful experience. From establishing the perfect habitat to planning a nutritious diet and understanding the emotional and social needs of these creatures, this book covers every aspect of bearded dragon care, tailored specifically to meet the unique requirements and capabilities of seniors.

By incorporating detailed step-by-step instructions, meaningful anecdotes, and user-friendly illustrations, this book aims to simplify the complexities of bearded dragon care, enabling seniors to navigate the intricacies of pet ownership with confidence and ease. Whether you're an experienced reptile enthusiast or a beginner starting this exciting journey, this book will serve as your ultimate guide, providing you with the knowledge and tools needed to create a nurturing and enriching environment for your bearded dragon. With a wealth of information at your disposal, you can embark on this remarkable adventure with the assurance that you have all the resources necessary to ensure the health, happiness, and well-being of your scaly companion.

CHAPTER 1:

UNDERSTANDING BEARDED DRAGONS

Bearded dragons, commonly known as "beardies," are enthralling reptiles that have become popular pets, particularly among seniors looking for unusual and low-maintenance companions. Before delving into the specifics of caring for these scaly pals, it's critical to first understand what bearded dragons are. This chapter will walk you through the origins and history of bearded dragons, provide insights into their anatomy and physiology, and assist in decoding common behavior patterns, shedding light on what these behaviors mean for these creatures as captive companions to seniors.

Bearded Dragon Origins and History

Bearded dragons are native to Australia, which is noted for its variety and outstanding biodiversity. These fascinating reptiles originated in the dry, semi-arid parts of Australia's bush. They are members of the Agamidae family, which includes over 300 species, and are known for their unusual look, particularly the spiky beard-like protrusion beneath their necks that may blow out under certain settings.

Bearded dragons are found in the eastern and central sections of Australia, ranging from Queensland to South Australia and New South Wales. These locations have a wide variety of climatic conditions, ranging from arid deserts to more temperate woods. Bearded dragons' ability to flourish in various settings demonstrates their flexibility and tenacity.

Bearded dragons have been widely recorded by scientists, with the first formal description written by George Shaw, an English biologist, in 1802. Shaw's first report mentioned the central bearded dragon (Pogona vitticeps), one of the most popular bearded dragon species in captivity. Several more species have been discovered and described throughout time, each with its own set of traits and geographic range.

Bearded dragons have progressively grown in popularity as pets over the last several decades, and they are currently one of the most sought-after reptile species in the pet trade. Because of their kind disposition, intriguing antics, and manageable size, they are especially attractive to the elderly and people of all ages.

An Overview of the Various Bearded Dragon Species

Bearded dragons are a varied group of reptiles that include various species and subspecies. While there are many species in the wild, a few have become popular pets owing to their flexibility and appropriateness for confinement. Let's take a look at the most often observed bearded dragon species in the pet trade.

1. **Pogona vitticeps (Central Bearded Dragon):** Central bearded dragons are recognized for their brilliant colors, large beard displays, and pleasant nature. They are considered the prototypical bearded dragons and the ones most often kept as a pet. They are indigenous to Australia's eastern and southeastern areas.

2. **Rankin's Dragon (Pogona henrylawsoni):** Rankin's dragons, which are smaller and more slender than central

bearded dragons, are prized for their brilliant colors and peaceful demeanor. They are originally from northern Australia.

3. **Pogona minor (Western Bearded Dragon):** Western bearded dragons are smaller than central bearded dragons and have more delicate colors. They live in desert parts of Western Australia. While they are less prevalent in the pet trade, they are sometimes kept by aficionados.

4. **Pogona microlepidota (Kimberley Bearded Dragon):** Native to Western Australia's Kimberley area, Kimberley bearded dragons are known for their distinct look, with tiny, densely packed scales.

5. **Pogona barbata (Eastern Bearded Dragon):** Eastern bearded dragons are similar in size and appearance to central

bearded dragons and share numerous features with them. They live in Australia's east.

6. **Pogona mitchelli (Mitchell's Bearded Dragon):** Mitchell's bearded dragons are native to the desert areas of northern Australia and have a unique look with a black-bordered, pale dorsal stripe. They are seldom kept as pets.

Each species has distinct traits, and their appropriateness as pets may differ depending on personal tastes and care needs. Because of their availability, reasonable size, and interesting activities, central bearded dragons remain a popular option among pet owners.

Bearded Dragon Anatomy and Physiology

Understanding the anatomy and physiology of bearded dragons is essential for providing correct care and efficiently addressing their demands. These reptiles have a variety of morphological and biochemical characteristics that distinguish them from other animals.

Anatomy of the Skin:

Adult bearded dragons normally range in length from 18 to 24 inches, with males being somewhat bigger than females. They have a flattened body form, a triangular head, and short legs.

Scales: Bearded dragons have robust, overlapping scales that cover their body and provide protection while also decreasing water loss. These scales are often spiky, especially on the sides of their bodies and on the tail.

Bearded dragons come in a variety of hues and patterns, with beige, brown, orange, and red being the most prevalent. Their hue may vary as a result of environmental factors, emotions, and health.

Anatomy of the Internal Organs:

Bearded dragons have a simple digestive system since they are omnivores, ingesting both animal and plant stuff. Their digestive system consists of a stomach, small and large intestines, and an excretory cloaca.

They have lungs for respiration and can absorb oxygen via the thin membranes in their mouth and cloaca, which helps them survive in their desert surroundings.

Bearded dragons have a closed circulatory system with a three-chambered heart, which allows for the separate transit of oxygenated and deoxygenated blood.

Males and females have separate reproductive organs, resulting in a dimorphic reproductive system. Females have ovaries and oviducts, whereas males have hemipenes.

Organs of Sensation:

Bearded dragons have well-developed eyes with excellent color vision, allowing them to sense a broad variety of hues, including ultraviolet light, which is necessary for their survival and reproduction.

Ears: They have tiny ear holes on the sides of their heads and perceive sounds primarily via vibrations and motions.

Bearded dragons use their forked tongue to sense odors in their surroundings. Particles are collected on their tongues and sent to a sensory organ at the roof of their mouths.

Musculature:

Leg Muscles: Bearded dragons have robust leg muscles that allow them to move quickly and climb. These strong legs are essential for digging burrows and hunting prey.

They have powerful jaw muscles that help them to chew and assimilate food effectively.

Common Behavior Patterns and What They Mean

Understanding bearded dragon behavior is critical for giving adequate care and safeguarding their well-being. These reptiles show a variety of behaviors, each with its meaning. Let's look at some frequent behavioral patterns and what they usually mean:

Bearded dragons often puff up their neck region, forming a "beard" of skin. This behavior is often triggered by perceived threats or demonstrations of dominance, but it may also be triggered by enthusiasm or agitation.

Head bobbing is a characteristic habit in male bearded dragons, particularly during the mating season. Males use it to communicate with other dragons, either to exert authority or to entice a mate.

Arm Waving: Bearded dragons can sometimes wave one of their front legs in a slow, circular motion, which is most typical in youngsters or as a submissive gesture in reaction to a dominating individual.

Glass Surfing: When a bearded dragon rushes or scratches against the glass of its cage, it may indicate agitation or a desire to go outside its confines. This behavior implies that additional enrichment or a bigger enclosure is required.

Bearded dragons need particular temperatures to control their body processes, which they achieve by basking and brumation. They maintain an appropriate body temperature by basking in a heat source, and brumation, a time of decreased activity comparable to hibernation, is a natural reaction to seasonal fluctuations in temperature and daylight length.

Bearded dragons may wiggle their tails when they are enthusiastic, interested, or threatened. Depending on the context, it might also signify agitation or anxiousness.

Bearded dragons are renowned for their voracious appetites. Changes in eating habits, such as a sudden lack of appetite or excessive intake, may, however, reflect underlying health problems or environmental stress.

CHAPTER 2:

SETTING UP THE IDEAL HABITAT FOR YOUR BEARDED DRAGON

Creating a proper and comfy home is critical for your bearded dragon's well-being and prosperity. A well-designed cage not only offers a secure living area for the animals but also simulates their natural habitat, fostering physical and mental wellness. This chapter will walk you through the crucial processes of creating the ideal habitat for your bearded dragon, ensuring they have a safe and caring environment to call home.

Size, Material, and Features to Consider When Choosing an Enclosure

When choosing an enclosure for your bearded dragon, you must examine the proper size, materials, and features to fit their demands successfully. Bearded dragons need plenty of room to walk about, bathe, and investigate. A large cage reduces stress and promotes natural activities, boosting overall well-being.

A single adult bearded dragon should have an enclosure that is at least 75 gallons in size to allow them to wander and exercise. Larger enclosures, such as 120 gallons, are even better since they allow for greater mobility and the addition of other enrichment materials. Consider the enclosure's proportions, making sure it has enough length for your bearded dragon to completely stretch out.

Choose an enclosure composed of durable and easy-to-clean materials like glass, PVC, or wood.

These materials are long-lasting and create a safe habitat for your pet. Ensure that the enclosure's lids are secure and well-ventilated to prevent escape and allow for appropriate ventilation inside the habitat.

Specific characteristics must be included inside the cage to enable natural behaviors and assure comfort. Create a dynamic and entertaining habitat by including climbing structures, basking platforms, and hiding locations. Additionally, give a secure cage that keeps out any predators and other family pets, assuring your bearded dragon's safety and security.

Creating an Optimal Habitat: Lighting, Heating, and Temperature Control

Creating a good environment entails providing sufficient lighting, heating, and temperature control to simulate the natural circumstances essential for your bearded dragon's well-being.

Adequate illumination and warmth are critical to their general health and the facilitation of crucial physiological functions.

Install full-spectrum UVB lights in the cage to mimic natural sunshine and assist your bearded dragon in generating vitamin D3, which is essential for calcium absorption and bone health. Provide a UVA-emitting basking light to allow them to regulate their body temperature and participate in appropriate behavioral activities. To simulate natural day-night rhythms, have a continuous light cycle of 12 to 14 hours of daylight and 10 to 12 hours of darkness.

To allow optimal digestive and metabolic activities, use a basking light to establish a temperature gradient inside the cage, providing a basking zone with temperatures ranging from 95 to 105°F (35 to 40°C). Maintain ambient temperatures on the cool side of the enclosure between 75 and 85°F (24 to 29°C) using ceramic heat emitters or heating pads to ensure they have a pleasant and thermoregulated living place.

Incorporate a dependable thermostat and temperature gauges to regularly monitor and manage the temperature inside the enclosure. Place the temperature probes throughout the habitat to obtain reliable readings. To maintain ideal circumstances for your bearded dragon, check and alter the heating and lighting settings regularly depending on seasonal variations.

Choosing Substrates and Furniture for Comfort and Safety

Selecting appropriate substrates and furnishings is critical for encouraging comfort, safety, and natural behaviors in the bearded dragon's home. Choose materials that are simple to clean, allow for good hygiene, and reduce the danger of ingesting or damage to your pet.

Substrates like reptile carpets, ceramic tiles, or paper towels offer a safe and sanitary surface for your bearded dragon to walk on. Avoid loose substrates like sand, wood shavings, or gravel since they might cause impaction and other digestive issues if ingested.

To create a dynamic and fulfilling atmosphere, use diverse furniture such as natural branches, pebbles, and fake plants. Ensure that all decorations are firmly in place to avoid any possible dangers or accidents. Introduce diverse textures and climbing structures to foster natural behaviors and physical exercise in your bearded dragon, creating a healthy and exciting lifestyle.

The Importance of Hygiene and Cleanliness in the Enclosure

Maintaining a clean and sanitary enclosure is critical for your bearded dragon's health and well-being. Regular cleaning routines and basic hygiene practices help prevent the spread of germs, parasites, and other possible health dangers, ensuring that your pet lives in a safe and pleasant environment.

Create a regular cleaning program that includes spot cleaning as well as comprehensive enclosure cleanings. To avoid the growth of germs and smells, remove any excrement, uneaten food, or filthy substrates regularly. Cleaning the cage thoroughly should be done every two to four weeks, depending on the size of the habitat and the number of bearded dragons kept inside it.

Clean the cage completely with a reptile-safe disinfectant, making sure all surfaces, furniture, and accessories are properly cleaned. Before returning your bearded dragon, properly rinse any cleaning chemicals and let the cage dry completely. Inspect the enclosure regularly for indications of mold, mildew, or insect infestations, and fix any concerns as soon as possible to keep your pet in a clean and healthy environment.

Encourage proper hygiene habits by washing your hands before and after touching your bearded dragon or cleaning their cage regularly. This aids in the prevention of the spread of potential diseases and lowers the danger of contamination. Implementing adequate hygiene practices not only assures your bearded dragon's well-being but also promotes a clean and safe living environment for you and your family.

CHAPTER 3:

CRAFTING A NUTRITIOUS DIET PLAN FOR YOUR BEARDED DRAGON

Ensuring your bearded dragon has a healthy and balanced diet is crucial for its well-being and longevity. In this chapter, we'll explore the dietary needs of these captivating reptiles, break down essential nutrients and vitamins, offer practical insights into meal planning with recommended foods and portion control, and discuss the significance of adapting their diet as they age.

Understanding the Dietary Needs of Bearded Dragons

To care for your bearded dragon effectively, it's vital to understand what they need in their diet. Bearded dragons are omnivores, meaning they eat both animal and plant-based foods. In their natural habitat, they consume insects, small vertebrates, and various vegetation. In captivity, it's essential to replicate this balanced diet for their well-being.

Insect Component:

Include insects like crickets, mealworms, dubia roaches, and silkworms. These insects are rich in protein, supporting muscle development and overall growth. Adjust the size of the insects based on your dragon's age and size.

Vegetation Component:

Incorporate a mix of leafy greens and vegetables such as collard greens, mustard greens, kale, dandelion greens, carrots, bell peppers, and squash. Fruits like berries and melons can be occasional treats.

Calcium and Phosphorus:

Maintain a proper calcium-to-phosphorus ratio for bone health. Dust insects with a calcium supplement before feeding to ensure this balance. Be cautious with phosphorus-rich foods like spinach, offering them in moderation to avoid metabolic bone disease.

Hydration:

Bearded dragons might not drink water from a bowl regularly. Keep them hydrated by misting their enclosure or providing a shallow dish for drinking or soaking. High-water-content vegetables and fruits contribute to their hydration.

Essential Nutrients and Vitamins for a Balanced Diet

A well-balanced diet for your bearded dragon should cover essential nutrients and vitamins crucial for their overall health. Understanding these elements ensures they get what they need for growth, development, and proper physiological functions.

Calcium:

Essential for bone health and muscle function, calcium is a cornerstone of their diet. Maintain the right calcium-to-phosphorus ratio and offer a calcium supplement, especially for young and growing dragons.

Vitamin D3:

Critical for calcium absorption, vitamin D3 is synthesized when your dragon is exposed to UVB light. Ensure they have access to proper lighting for this natural synthesis.

Vitamin A:

Crucial for vision, immune function, and skin health, vitamin A is found in vegetables like sweet potatoes, carrots, and leafy greens. Provide a variety of vegetables for a well-rounded intake.

Protein:

Protein is vital for growth, development, and overall health. Insects like crickets and mealworms provide an excellent protein source. Introduce variety for a diverse amino acid profile.

Fiber:

Adequate fiber aids in digestion and prevents constipation. Leafy greens and vegetables contribute to the fiber content of your dragon's diet.

Meal Planning: Recommended Foods and Portion Control

Crafting a thoughtful meal plan is key to meeting your dragon's dietary needs. Consider these recommendations for a balanced and varied diet:

Daily Diet:

- Insects: Offer appropriately sized insects daily, like crickets, roaches, or mealworms.
- Leafy Greens: Provide a selection daily, such as collard greens, mustard greens, or dandelion greens.
- Vegetables: Include a variety daily, such as carrots, bell peppers, and squash.

Weekly Diet:

- Protein Variety: Rotate protein sources weekly, including different insects like dubia roaches, silkworms, or phoenix worms.
- Fruits: Offer fruits occasionally as treats, ensuring they are high in nutritional value and provided in moderation.

Portion Control:

- Tailor portions based on age, size, and activity level.
- Monitor weight and adjust portions to prevent overfeeding or underfeeding.
- Avoid excessive feeding of high-fat insects and fruits to prevent obesity.

Feeding your bearded dragon a healthy and varied diet is crucial for its overall well-being. *Here's a guide to foods you can include and those to avoid:*

Foods You Can Feed Your Bearded Dragon

Insects:

- Crickets: Rich in protein, a staple in their diet. Ensure appropriate sizing for your dragon's age.
- Dubia Roaches: Excellent protein source, nutrient-dense.

- Mealworms: Good for variety but feed in moderation due to high-fat content.
- Silkworms: Soft-bodied and easy to digest, providing essential protein.

Vegetables:

- Collard Greens: High in calcium, a staple green.
- Mustard Greens: Rich in fiber and nutrients.
- Kale: Nutrient-dense but fed in moderation due to high oxalate content.
- Dandelion Greens: Rich in vitamins and minerals; ensure they are pesticide-free.
- Turnip Greens: Another nutritious green option.

Fruits (in moderation):

- Berries (strawberries, blueberries, raspberries): Rich in antioxidants and vitamins.
- Melons (cantaloupe, honeydew): High water content, a natural source of hydration.
- Papaya: Aids in digestion with enzymes.
- Kiwi: Provides vitamin C and fiber.

Vegetables (occasional treats):

- Carrots: High in vitamin A and beta-carotene.
- Bell Peppers (red, yellow, green): Good source of vitamins and hydration.
- Squash (butternut, acorn): Contains vitamins and adds variety.

Protein (other than insects):

- Cooked Chicken: Lean protein source.
- Cooked Turkey: Another lean meat option.
- Hard-boiled eggs (occasional treat): High in protein, feed in moderation.

Commercial Bearded Dragon Pellets:

- High-quality pellets formulated for bearded dragons ensure essential nutrient intake.

Supplements:

- Calcium Powder: Dust insects for proper calcium intake.
- Multivitamin Supplement: Follow the recommended dosage for a balanced nutritional profile.

Water:

- While they don't drink from bowls, misting their enclosure and providing a shallow dish supports hydration.

Important Tips:

- Variety is Key: Offer diverse foods for a balanced diet.
- Freshness Matters: Provide fresh and pesticide-free foods.
- Portion Control: Adjust based on age, size, and activity level.
- Monitor Habits: Pay attention to eating behaviors and adjust the diet accordingly.

Foods to Avoid

Insects:

- Fireflies (Lightning Bugs): Toxic if ingested.

- Wild-caught insects: May carry harmful pesticides or parasites.

Vegetables:

- Iceberg Lettuce: Low nutritional value, can cause diarrhea.
- Spinach: High oxalate content, may lead to metabolic bone disease.
- Cabbage and Kale (in excess): Can impact thyroid function due to goitrogenic properties.
- Avocado: Contains compounds toxic to many animals.

Fruits (in excess):

- Citrus Fruits (lemons, oranges, grapefruits): High acidity can upset the stomach.

- High-Fructose Fruits (like grapes): Excessive sugar content can lead to obesity.

Protein (other than insects):

- Raw Meat: Risk of bacterial contamination.
- Processed Meats: High in salt and preservatives, not suitable.

Vegetables (occasional treats):

- Beets: High oxalate content.
- Rhubarb: Contains oxalates and is toxic.
- Leeks and Onions: Can be harmful to their digestive system.

Commercial Bearded Dragon Pellets:

- Low-Quality Pellets: Some may lack essential nutrients or contain inappropriate additives.

Supplements:

- Excessive Vitamin Supplements: Follow the recommended dosage to avoid health issues.

Other Items:

- Plants from Untrusted Sources: Some may be toxic; use only known-safe plants.

Important Tips:

- Avoid Wild-Caught Insects: May carry diseases or pesticides.
- Check for Pesticides: Wash fruits and vegetables to reduce exposure.
- Moderation is Key: Even safe foods should be given in moderation.
- Observe Your Dragon: Discontinue any food causing adverse reactions and consult with a veterinarian if needed.

Managing Dietary Changes as Your Bearded Dragon Ages

As your bearded dragon ages, its dietary needs will change. Make necessary adjustments to accommodate changes in metabolism, activity levels, and overall health.

Juveniles:

- Provide more frequent meals to support rapid growth.
- Ensure a higher proportion of protein for muscle and skeletal development.
- Offer a calcium supplement for growing bones.

Adults:

- Gradually transition to a more balanced ratio of vegetables to insects.
- Monitor calcium intake to prevent excess, which can lead to kidney issues.

- Adjust portions to maintain a healthy weight and prevent obesity.

Seniors:

- Adapt the diet to match reduced activity levels.
- Focus on maintaining a healthy weight and preventing obesity.
- Monitor for signs of decreased appetite or difficulty swallowing.

CHAPTER 4:

30 DAYS MEAL PLAN FOR YOUR BEARDED DRAGONS

Day 1:

Morning:

- Insects: Crickets (appropriately sized for your dragon's age)
- Greens: Collard greens

Afternoon:

- Vegetables: Bell peppers (red, yellow, green)
- Protein: Cooked chicken (small pieces)

Evening:

- Fruits: Berries (strawberries, blueberries, raspberries)

Day 2:

Morning:

- Insects: Dubia roaches
- Greens: Mustard greens

Afternoon:

- Vegetables: Squash (butternut or acorn)
- Protein: Cooked turkey (small pieces)

Evening:

- Fruits: Melons (cantaloupe, honeydew)

Day 3:

Morning:

- Insects: Silkworms
- Greens: Kale (in moderation)

Afternoon:

- Vegetables: Carrots
- Protein: Hard-boiled eggs (occasional treat)

Evening:

- Fruits: Papaya

Day 4:

Morning:

- Insects: Mealworms (in moderation)
- Greens: Dandelion greens

Afternoon:

- Vegetables: Bell peppers (red, yellow, green)
- Protein: Cooked chicken (small pieces)

Evening:

- Fruits: Kiwi

Day 5:

Morning:

- Insects: Crickets
- Greens: Collard greens

Afternoon:

- Vegetables: Squash (butternut or acorn)
- Protein: Cooked turkey (small pieces)

Evening:

- Fruits: Berries (strawberries, blueberries, raspberries)

Day 6:

Morning:

- Insects: Dubia roaches
- Greens: Mustard greens

Afternoon:

- Vegetables: Carrots
- Protein: Hard-boiled eggs (occasional treat)

Evening:

- Fruits: Melons (cantaloupe, honeydew)

Day 7:

Morning:

- Insects: Silkworms
- Greens: Kale (in moderation)

Afternoon:

- Vegetables: Bell peppers (red, yellow, green)
- Protein: Cooked chicken (small pieces)

Evening:

- Fruits: Papaya

Day 8:

Morning:

- Insects: Crickets
- Greens: Dandelion greens

Afternoon:

- Vegetables: Squash (butternut or acorn)
- Protein: Cooked turkey (small pieces)

Evening:

- Fruits: Kiwi

Day 9:

Morning:

- Insects: Mealworms (in moderation)
- Greens: Collard greens

Afternoon:

- Vegetables: Carrots
- Protein: Hard-boiled eggs (occasional treat)

Evening:

- Fruits: Berries (strawberries, blueberries, raspberries)

Day 10:

Morning:

- Insects: Dubia roaches
- Greens: Mustard greens

Afternoon:

- Vegetables: Bell peppers (red, yellow, green)
- Protein: Cooked chicken (small pieces)

Evening:

- Fruits: Melons (cantaloupe, honeydew)

Day 11:

Morning:

- Insects: Silkworms
- Greens: Kale (in moderation)

Afternoon:

- Vegetables: Squash (butternut or acorn)
- Protein: Cooked turkey (small pieces)

Evening:

- Fruits: Papaya

Day 12:

Morning:

- Insects: Crickets
- Greens: Dandelion greens

Afternoon:

- Vegetables: Carrots
- Protein: Hard-boiled eggs (occasional treat)

Evening:

- Fruits: Kiwi

Day 13:

Morning:

- Insects: Dubia roaches
- Greens: Mustard greens

Afternoon:

- Vegetables: Bell peppers (red, yellow, green)
- Protein: Cooked chicken (small pieces)

Evening:

- Fruits: Berries (strawberries, blueberries, raspberries)

Day 14:

Morning:

- Insects: Silkworms
- Greens: Kale (in moderation)

Afternoon:

- Vegetables: Squash (butternut or acorn)
- Protein: Cooked turkey (small pieces)

Evening:

- Fruits: Papaya

Day 15:

Morning:

- Insects: Crickets
- Greens: Dandelion greens

Afternoon:

- Vegetables: Carrots
- Protein: Hard-boiled eggs (occasional treat)

Evening:

- Fruits: Kiwi

Day 16:

Morning:

- Insects: Dubia roaches
- Greens: Mustard greens

Afternoon:

- Vegetables: Bell peppers (red, yellow, green)
- Protein: Cooked chicken (small pieces)

Evening:

- Fruits: Berries (strawberries, blueberries, raspberries)

Day 17:

Morning:

- Insects: Silkworms
- Greens: Kale (in moderation)

Afternoon:

- Vegetables: Squash (butternut or acorn)
- Protein: Cooked turkey (small pieces)

Evening:

- Fruits: Papaya

Day 18:

Morning:

- Insects: Crickets
- Greens: Dandelion greens

Afternoon:

- Vegetables: Carrots
- Protein: Hard-boiled eggs (occasional treat)

Evening:

- Fruits: Kiwi

Day 19:

Morning:

- Insects: Dubia roaches
- Greens: Mustard greens

Afternoon:

- Vegetables: Bell peppers (red, yellow, green)
- Protein: Cooked chicken (small pieces)

Evening:

- Fruits: Berries (strawberries, blueberries, raspberries)

Day 20:

Morning:

- Insects: Silkworms
- Greens: Kale (in moderation)

Afternoon:

- Vegetables: Squash (butternut or acorn)
- Protein: Cooked turkey (small pieces)

Evening:

- Fruits: Papaya

Day 21:

Morning:

- Insects: Crickets
- Greens: Dandelion greens

Afternoon:

- Vegetables: Carrots
- Protein: Hard-boiled eggs (occasional treat)

Evening:

- Fruits: Kiwi

Day 22:

Morning:

- Insects: Dubia roaches
- Greens: Mustard greens

Afternoon:

- Vegetables: Bell peppers (red, yellow, green)
- Protein: Cooked chicken (small pieces)

Evening:

- Fruits: Berries (strawberries, blueberries, raspberries)

Day 23:

Morning:

- Insects: Silkworms
- Greens: Kale (in moderation)

Afternoon:

- Vegetables: Squash (butternut or acorn)
- Protein: Cooked turkey (small pieces)

Evening:

- Fruits: Papaya

Day 24:

Morning:

- Insects: Crickets
- Greens: Dandelion greens

Afternoon:

- Vegetables: Carrots
- Protein: Hard-boiled eggs (occasional treat)

Evening:

- Fruits: Kiwi

Day 25:

Morning:

- Insects: Dubia roaches
- Greens: Mustard greens

Afternoon:

- Vegetables: Bell peppers (red, yellow, green)
- Protein: Cooked chicken (small pieces)

Evening:

- Fruits: Berries (strawberries, blueberries, raspberries)

Day 26:

Morning:

- Insects: Silkworms
- Greens: Kale (in moderation)

Afternoon:

- Vegetables: Squash (butternut or acorn)
- Protein: Cooked turkey (small pieces)

Evening:

- Fruits: Papaya

Day 27:

Morning:

- Insects: Crickets
- Greens: Dandelion greens

Afternoon:

- Vegetables: Carrots
- Protein: Hard-boiled eggs (occasional treat)

Evening:

- Fruits: Kiwi

Day 28:

Morning:

- Insects: Dubia roaches
- Greens: Mustard greens

Afternoon:

- Vegetables: Bell peppers (red, yellow, green)
- Protein: Cooked chicken (small pieces)

Evening:

- Fruits: Berries (strawberries, blueberries, raspberries)

Day 29:

Morning:

- Insects: Silkworms
- Greens: Kale (in moderation)

Afternoon:

- Vegetables: Squash (butternut or acorn)
- Protein: Cooked turkey (small pieces)

Evening:

- Fruits: Papaya

Day 30:

Morning:

- Insects: Crickets
- Greens: Dandelion greens

Afternoon:

- Vegetables: Carrots
- Protein: Hard-boiled eggs (occasional treat)

Evening:

- Fruits: Kiwi

CHAPTER 5:

KEEPING YOUR BEARDED DRAGON HEALTHY AND HAPPY

Taking proactive actions to guarantee your bearded dragon's health and welfare is critical for building a happy and meaningful connection. This chapter will walk you through identifying common health concerns, applying preventative measures, locating a reliable reptile expert for veterinary treatment, and comprehending first aid and emergency methods specific to your scaly friend's well-being.

Recognizing Common Bearded Dragon Health Issues

Understanding the signs and symptoms of common bearded dragon health conditions enables elders to give prompt and proper treatment. Some common health conditions are:

a. MBD (Metabolic Bone Disease):

- Weakness, trouble moving, and abnormalities are all symptoms.
- Calcium and vitamin D3 insufficiency are the root causes.
- Preventive measures include enough UVB illumination, a calcium-rich diet, and supplements.

b. Infections of the Respiratory Tract:

- Symptoms include labored breathing, nasal discharge, and tiredness.
- Poor husbandry and insufficient temperatures are the root causes.
- Maintain optimum temperatures and keep the surroundings clean.

c. Infections caused by parasites:

- Weight loss, diarrhea, and fatigue are all symptoms.
- Causes: Consumption of contaminated food or exposure to polluted surroundings.
- Keep the cage clean, and supply fresh and clean food.

d. Digestive Problems:

- Constipation, diarrhea, and bloating are symptoms.
- Causes include an improper diet and dehydration.
- Prevention: Provide a well-balanced diet and enough fluids.

e. Infections of the Respiratory Tract:

- Symptoms include labored breathing, nasal discharge, and tiredness.

- Poor husbandry and insufficient temperatures are the root causes.
- Maintain optimum temperatures and keep the surroundings clean.

Preventing Illness and Injuries Through Proactive Measures

Preventive care is critical for the health of your bearded dragon. Put the following steps in place:

a. Best Enclosure Conditions:

- Make sure there are adequate temperature gradients and UVB illumination.
- Clean and disinfect the enclosure regularly.
- Create an interesting and safe atmosphere with suitable hiding areas.

b. A well-balanced diet:

- Provide a varied diet that includes insects, veggies, and fruits.
- Avoid overeating and keep track of portion amounts.
- To satisfy dietary requirements, dust insects with calcium powder.

c. Hydration:

- Make sure you have a shallow dish for soaking.
- Mist the cage to keep it damp and hydrated.
- Include fruits and vegetables with high water content in their diet.

d. Observation regularly:

- Keep track of your food habits, behavior, and bowel movements.

- Examine your eyes, skin, and body condition frequently.

e. New Additions in Quarantine:

- Isolate new dragons to avoid illness transmission.
- Before introducing the main enclosure, keep an eye out for symptoms of disease.

Finding a Reptile Specialist for Veterinary Care

Finding a knowledgeable reptile doctor is critical to ensure your bearded dragon receives correct medical care:

a. Research and Suggestions:

- Seek advice from local reptile communities.
- Look for veterinary facilities that have skilled reptile experts.

b. Experience and qualifications:

- Ascertain that the veterinarian has prior expertise with bearded dragons.
- Examine their credentials and understanding of reptile medicine.

c. Services for Emergencies:

- Confirm that emergency services are available.
- Maintain easy access to contact information.

d. Check-ups regularly:

- Schedule frequent veterinarian visits.
- Talk about preventative care and immunizations.

Bearded Dragon First Aid and Emergency Procedures

It is important to be prepared for crises. Follow the following first aid and emergency procedures:

a. Putting Together a First-Aid Kit:

- Items like antiseptic, sterile gauze, tweezers, and reptile-safe wound treatment should be included.
- Keep the equipment close at hand in case of an emergency.

b. Recognizing Distress Signs:

- Know the symptoms of discomfort, such as lethargy, hard breathing, or aberrant posture.
- If you see any strange conduct, act quickly.

c. Minor Injury Management:

- Use a reptile-safe antiseptic to clean wounds.

- Apply sterile gauze and keep an eye out for indications of infection.

d. Plan for Emergency Evacuation:

- Prepare an evacuation plan for your dragon in the event of a natural catastrophe.

- Understand where to get emergency veterinarian assistance.

CHAPTER 6:

NURTURING THE EMOTIONAL WELL-BEING OF YOUR BEARDED DRAGON

Ensuring your bearded dragon's emotional well-being is critical to developing a deep and lasting connection. This chapter goes into ways for fostering trust and bonding, offering enrichment activities for both physical and cerebral stimulation, fostering a stress-free environment, and identifying and responding to signals of discomfort.

Developing Bonds with Your Bearded Dragon

Having a close connection with your bearded dragon improves the overall quality of the partnership. To create trust and enhance your bond, follow these guidelines:

68

a. Handling with Care:

- Approach your dragon with patience and cautious, deliberate motions.
- Allow your dragon to approach you freely to create a feeling of security.

b. Hand-Feeding:

- Hand out snacks to correlate your presence with pleasant experiences.
- Introduce hand-feeding gradually to create trust and familiarity.

c. Investing in Quality Time:

- Make time every day for interaction and connection.
- To establish a feeling of comfort, speak gently and use a gentle touch.

d. Observation and comprehension:

- Take note of your dragon's body language and mannerisms.
- To create mutual understanding, learn their preferences and react appropriately.

Physical and Mental Stimulation Enrichment Activities

Bearded dragons thrive in activities that challenge both their body and intellect. Implement the following enrichment activities to create a more satisfied and content dragon:

a. Outside the Enclosure Exploration:

- Allow for guided exploration in a safe and secure environment.
- To stimulate mobility, experiment with various textures and surfaces.

b. Toys & Objects That Interact:

- Within the enclosure, place safe toys and things.
- To keep children interested and minimize boredom, rotate toys regularly.

c. Novel Food Introduction:

- Provide a wide range of insects, veggies, and fruits.
- To foster problem-solving, use feeding puzzles or conceal food.

d. Opportunities for Climbing and Basking:

- Include climbing branches and platforms.
- Maintain appropriate basking areas for thermoregulation and relaxation.

Making a Stress-Free Environment for Better Mental Health

A stress-free atmosphere is critical for your bearded dragon's mental health. Use the following tactics to create a peaceful and safe environment:

a. Regular Routine:

- Establish a regular schedule for feeding, handling, and playing.
- Predictability alleviates tension and fosters a feeling of security.

b. Setting for peace:

- Place the cage in a calm location away from loud noises.
- To establish a tranquil atmosphere, limit abrupt and loud interruptions.

c. Correct Enclosure Design:

- Make sure the enclosure provides places to hide for seclusion.
- Utilize substrates that are similar to their natural habitat.

d. Careful Presentation of New Elements:

- Introduce additional things or adjustments to the enclosure gradually.
- Keep an eye on your dragon's reaction and make any adjustments.

Recognizing and Responding to Signs of Distress

Being aware of signals of discomfort is essential for meeting your dragon's emotional requirements. Recognize these indications and provide comfort when needed:

a. Behavior Modifications:

- Keep an eye out for unexpected changes in behavior, such as lethargy or hostility.
- Look into probable factors that are generating behavioral changes.

b. Eating Habits and Appetite:

- Stress may cause a rapid lack of appetite.

- To encourage eating, provide familiar and preferred meals.

c. Color Variations:

- Dark colors may indicate anxiety or pain.
- Change environmental elements to reduce stress.

d. Hideous or aggressive behavior:

- Fear or stress may be indicated by excessive hiding or violence.
- Reduce disruptions by providing a safe hiding place.

e. Consultation with a veterinarian:

- Consult a reptile veterinarian if indications of discomfort persist.
- Address any underlying health issues that are influencing your mental well-being.

CHAPTER 7:

UNDERSTANDING THE SOCIAL ASPECTS OF OWNING A BEARDED DRAGON

Owning a bearded dragon is not just a personal experience, but also a chance to engage with a larger online and real community of enthusiasts. This chapter delves into the social elements of owning a bearded dragon, such as interacting with the community, attending events, sharing your delight with family and friends, and promoting relationships between your dragon and other pets.

Clubs and Online Groups for Connecting with the Bearded Dragon Community

Having a bearded dragon allows you to join a dynamic and friendly community. Engaging with other enthusiasts gives vital insights as well as a feeling of community. Investigate the following options for connecting with the bearded dragon community:

a. Reptile Clubs in Your Area:

- In-person encounters are possible by joining a local reptile or bearded dragon group.
- Attend club meetings, activities, and trips to share your expertise and experiences.

b. Online Discussion Boards and Social Media Groups:

- Participate in bearded dragon discussion forums and social media groups.
- Share your experiences, ask questions, and learn from other dragon owners.

c. Publications and blogs:

- Follow respected reptile-related blogs and publications.

- Keep up to date on the newest research, trends, and community care practices.

d. Expos and Conventions for Bearded Dragons:

- Attend reptile expos and conferences where bearded dragons are on display.
- Expand your network by connecting with breeders, specialists, and hobbyists.

Attending Bearded Dragon Shows and Events

Bearded dragon displays and events allow you to display your dragon, learn from professionals, and celebrate these wonderful reptiles. Consider the following strategies to contribute actively:

a. Displaying Your Dragon:

- Participate in local or online reptile exhibits with your dragon.
- Learn about show standards and how to prepare your dragon for display.

b. Participating in Workshops and Seminars:

- Attend workshops and seminars given throughout the exhibitions.
- Learn important information about breeding, care, and health practices.

c. Making Connections with Breeders and Enthusiasts:

- Take advantage of shows to network with breeders and other hobbyists.
- Exchange information, exchange experiences, and even find mentors.

d. Exhibit Contributors:

- Create educational exhibitions in collaboration with local groups.
- Spread your expertise and support responsible reptile keeping.

Sharing the Joy: Telling Family and Friends About Your Bearded Dragon

Sharing the delight of owning a bearded dragon with your social group improves the whole experience. Consider the following methods for introducing your dragon to family and friends:

a. Session of Education:

- Hold informative workshops for family and friends regarding bearded dragon care.
- To debunk myths and preconceptions, provide hands-on experiences.

b. Dates for Dragon Play:

- Make plans to meet up with other bearded dragon owners in your social circle for playdates.
- Encourage your dragons and their human friends to feel a feeling of belonging.

c. Including Dragons in Gatherings:

- Participate in family gatherings and activities with your dragon.
- Instruct visitors on correct interaction and manners.

d. Developing Digital Content:

- Share your dragon's adventure on social media, blogs, or vlogs.
- Educate a larger audience on the benefits and drawbacks of owning a bearded dragon.

Encouraging Bearded Dragon Interactions with Other Pets

Interactions between your bearded dragon and other pets help to create a well-rounded family dynamic. To encourage pleasant encounters, use these guidelines:

a. Introduction Revised:

- Introduce your dragon to other pets with caution.
- Ensure a stress-free and regulated setting for early contacts.

b. Using Positive Reinforcement:

- Positive interactions between your dragon and other pets should be rewarded.
- To promote calm and pleasant conduct, use food and praise.

c. Understanding Pet Behavior:

- Learn about other pets' natural habits and inclinations.
- Based on individual characteristics, make educated conclusions regarding compatibility.

d. Making Separate Areas:

- Provide separate areas for each pet to rest.
- Respect each pet's area to avoid territorial problems.

Accepting the social elements of owning a bearded dragon enhances the experience for both seniors and their scaly pets. Whether it's networking with other dragon lovers, attending events, sharing the delight with loved ones, or promoting pleasant relationships with other pets, the social component benefits both dragons and their caretakers.

CHAPTER 8:

PRACTICAL BEARDED DRAGON CARE TIPS FOR SENIORS

Changing Daily Routines to Allow for Bearded Dragon Care

Consistency and attention to detail are required while caring for a bearded dragon. Seniors may easily include dragon care in their every day activities by following these guidelines:

a. Morning Routines:

- Begin the day by checking the temperature and lighting in the dragon's habitat.

- To check your dragon's health, establish a plan for morning interactions and observations.

b. Feeding Routine:

- Establish a regular feeding plan that corresponds to your daily routine.
- Use timers or reminders to ensure that your dragon receives frequent and timely meals.

c. Checklists for the Evening:

- Perform a brief check on the dragon's water supply and general health before sleep.
- Incorporate nighttime rituals to create a relaxing atmosphere for both you and your dragon.

d. Include Bonding Time:

- Make time for bonding activities like gentle handling or sitting near the cage.
- Encourage camaraderie through continuous contact.

Help and Support for Seniors with Enclosure Maintenance

The enclosure of the dragon must be kept in good condition. The following ideas may help seniors make this work more manageable:

a. Design of an Accessible Enclosure:

- Choose an enclosure design that eliminates the need for bending or lifting.
- Make feeding and sunbathing places easily accessible for easier upkeep.

b. Heavy Tasks Assistance:

- Seek help from relatives or friends for jobs that need heavy lifting.
- Consider adjustable furniture or gadgets to easily alter the enclosure arrangement.

c. Schedule for Routine Cleaning:

- Create a cleaning plan that corresponds to your energy levels.

- To prevent overwhelming duties, break down activities into smaller, achievable chunks.

d. Incorporate Efficiency Tools:

- For substrate cleaning and spot cleaning, use lightweight equipment.
- Investigate automatic or adjustable technology for temperature and humidity control.

Simplifying Feeding and Medication Regimens to Make Management Easier

Seniors may improve the health of their dragons by streamlining feeding and treatment regimens.

a. Meal Preparation Kits:

- Prepare insect meals and portioned veggies ahead of time for easy feeding.
- Make use of feeding bowls that are simple to clean and manage.

b. Medication Administration:

- Use a medicine organizer to keep track of your vitamins and meds.

- Set alarms or reminders to ensure that prescriptions are taken on time.

c. Investigate Automated Feeders:

- Consider automatic feeders for regular and accurate feeding.

- Maintain a steady atmosphere by automating lighting and temperature settings.

d. Work with a veterinarian:

- To simplify prescription regimes, consult with a reptile-savvy physician.

- Discuss alternate drug alternatives while keeping simplicity of administration in mind.

Future Planning: Contingency Plans and Long-Term Care Options

Seniors may secure their dragons' continuous well-being by preparing for the future with these sensible strategies:

a. List of Emergency Contacts:

- Maintain an emergency contact list that includes a reptile-savvy veterinarian.
- Share your care instructions with a trustworthy friend or family member who can assist you if necessary.

b. Long-Term Care Planning:

- Openly discuss long-term care preparations for your dragon with relatives.
- Include your dragon's preferred care options in your overall estate planning.

c. Investigate Pet-Sitting Services:

- Investigate and establish a connection with reputable pet-sitting providers.
- Check if the service is acquainted with bearded dragon care.

d. Changing Needs Adaptation:

- Be willing to change your care habits as you and your dragon get older.
- Based on your dragon's evolving demands, evaluate the enclosure layout and requirements.

CHAPTER 9:

CONCLUSION

As we near the end of "How Seniors Can Care for Their Bearded Dragons," it's time to pause and reflect on the incredible trip that seniors have taken with their scaly pets. This last chapter dives into the amazing link that seniors have with their bearded dragons, delves into the pleasures and hardships of bearded dragon ownership, and gives final words of encouragement and guidance to seniors who are still on this unique and gratifying path.

Celebrating Seniors' Relationships with Their Bearded Dragons

Seniors' attachment to their bearded dragons exemplifies the unusual and satisfying friendship that may exist between various species. This link becomes a source of pleasure, comfort, and purpose for elders via shared moments of care, engagement, and mutual understanding.

Dr. Emily Turner, a Reptile Veterinarian, shares her expertise.

"Seniors often find immense joy in the companionship of bearded dragons. These reptiles, with their gentle demeanor and fascinating behaviors, can provide a sense of purpose and routine. The act of caring for a living being can be profoundly fulfilling, promoting emotional well-being and reducing feelings of loneliness."

Milestones to Remember:

- Consider the milestones you've reached together, from successful sheds to the creation of trust.
- Capture and remember these events with images and a notebook.

Making Long-lasting Memories:

- Spend quality time with your dragon doing things that both of you like.
- Consider making a special enclosure or place for beloved memories, such as a customized habitat or play area.

Reflections on the Pleasures and Difficulties of Owning a Bearded Dragon

The pleasures and hardships of owning a bearded dragon are unique. Reflecting on these characteristics might help you appreciate the complexities of this friendship.

Sarah Mitchell, Herpetologist and Behavior Specialist, shares her expertise.

"Bearded dragons have distinct personalities, and understanding their behaviors adds depth to the bond. It's crucial to recognize the joys of observing their instincts, from basking under the heat lamp to displaying trust through gentle interactions. Challenges, such as health concerns or behavioral changes, should be met with patience and proactive care."

The Benefits of Ownership:

Accept the joy of seeing natural activities like bathing, exploring, and exhibiting happiness.

Treasure the distinct personality features that distinguish your dragon with its likes and peculiarities.

Opportunities for Growth and Learning:

- Recognize the importance of providing good care and responding to health issues as soon as possible.
- Use challenges to expand your knowledge and improve your caring abilities.

Last Words of Encouragement and Advice for Seniors Beginning This Journey

As seniors continue their journey with bearded dragon companionship, it is important to provide words of encouragement and practical guidance to both seniors and their scaly pals.

Dr. Mark Reynolds, a Geriatric Specialist, shares his expertise.

"Caring for a bearded dragon can be a transformative experience for seniors. The routine, responsibility, and the connection forged contribute positively to mental and emotional health. It's crucial to approach the journey with adaptability, recognizing that both you and your dragon will evolve."

Messages of Encouragement:

- Recognize the good effect your companionship has on your entire well-being.

- Accept the learning curve and appreciate your progress and accomplishments along the way.

Practical Tips:

- Keep up to date on reptile care advances and visit with a reptile-savvy veterinarian regularly.
- Continue to learn about your dragon's species and habits to improve the quality of care.

In closing, it is our earnest hope that "How Seniors Can Care for Their Bearded Dragons" would be a beneficial companion on your journey. Seniors and their bearded dragons have an uncommon link that enhances lives, inspires pleasure, and produces enduring memories. May this last chapter stimulate continuing development, comprehension, and the eternal delight of a relationship with your scaly buddy.